NATURE CLOSE-UP

CRICKETS AND GRASSHOPPERS

TEXT BY ELAINE PASCOE

PHOTOGRAPHS BY DWIGHT KUHN

BLACKBIRCH PRESS, INC.

WOODBRIDGE, CONNECTICUT

Published by Blackbirch Press, Inc.
260 Amity Road
Woodbridge, CT 06525

To my daughter, Michele
–D.K.

To Bron
–E.P.

First Edition

Email: staff@blackbirch.com
Web site: www.blackbirch.com

Printed in the United States

10 9 8 7 6 5 4 3 2 1

front cover: grasshopper on a flower
back cover: (left to right) egg plug of a grasshopper, the nymph stage of a grasshopper, a grasshopper's discarded skin, adult two-striped grasshopper

Library of Congress Cataloging-in-Publication Data

Pascoe, Elaine.
Crickets and grasshoppers / by Elaine Pascoe. — 1st ed.
p. cm. — (Nature close-up)
Includes bibliographical references (p.) and index.
Summary: Describes the physical characteristics, habitats, and life cycle of crickets and grasshoppers. Includes related activities.
ISBN 1-56711-176-9 (lib. binding : alk. paper)
1. Crickets—Juvenile literature. 2. Grasshoppers—Juvenile literature. 3. Crickets—Experiments—Juvenile literature. 4. Grasshoppers—Experiments—Juvenile literature. [1. Crickets. 2. Grasshoppers. 3. Crickets—Experiments. 4. Grasshoppers—Experiments. 5. Experiments.] I. Title. II. Series: Pascoe, Elaine. Nature close-up.
QL508.G8P37 1999 97-43572
595.7'26—dc21 CIP
AC

Note on metric conversions: The metric conversions given in Chapters 2 and 3 of this book are not always exact equivalents of U.S. measures. Instead, they provide a workable quantity for each experiment in metric units. The abbreviations used are:

cm	centimeter	**kg**	kilogram
m	meter	**l**	liter
g	gram	**cc**	cubic centimeter

CONTENTS

1

Musical Insects

A buzzing *zeee-zeee-zeee* rises from meadows on warm summer days. As evening falls, *chir-r-ip chir-r-ip* and *treet-treet-treet* ring from trees and fields. Summer is filled with music performed by an insect orchestra made up of grasshoppers and crickets.

There are many different kinds of grasshoppers and crickets, found all over the world. A few of them are serious pests—especially the migrating grasshoppers known as locusts, which travel in huge swarms and do great damage to crops. But most crickets and grasshoppers do little or no harm and are fascinating to watch and to hear. Crickets and grasshoppers have no voices. They produce their calls by rubbing body parts together, as if they were playing fiddles.

Camel cricket

MEMBERS OF THE FAMILY

Crickets live in fields, rock walls, trees, underground, and in houses. There are many types, some only 3/8 inch (about 1 cm) long and others well over an inch (2.5 cm) in length. Field crickets are mostly black and brown. They are common in meadows and backyards, and they're often found indoors as well. True house crickets are lighter in color. Originally house crickets did not live in North America. They were accidentally carried over from Europe by early settlers.

Tree and bush crickets, mostly pale greenish-yellow and white, are heard more often than seen because they blend in so well with their habitats. The oddest crickets are the big mole crickets, which dig underground burrows. Males often sit in the doorways of their burrows and chirp loudly at night. Evening is the time when you will hear crickets the most—it is when all crickets are most active.

Field cricket

House cricket

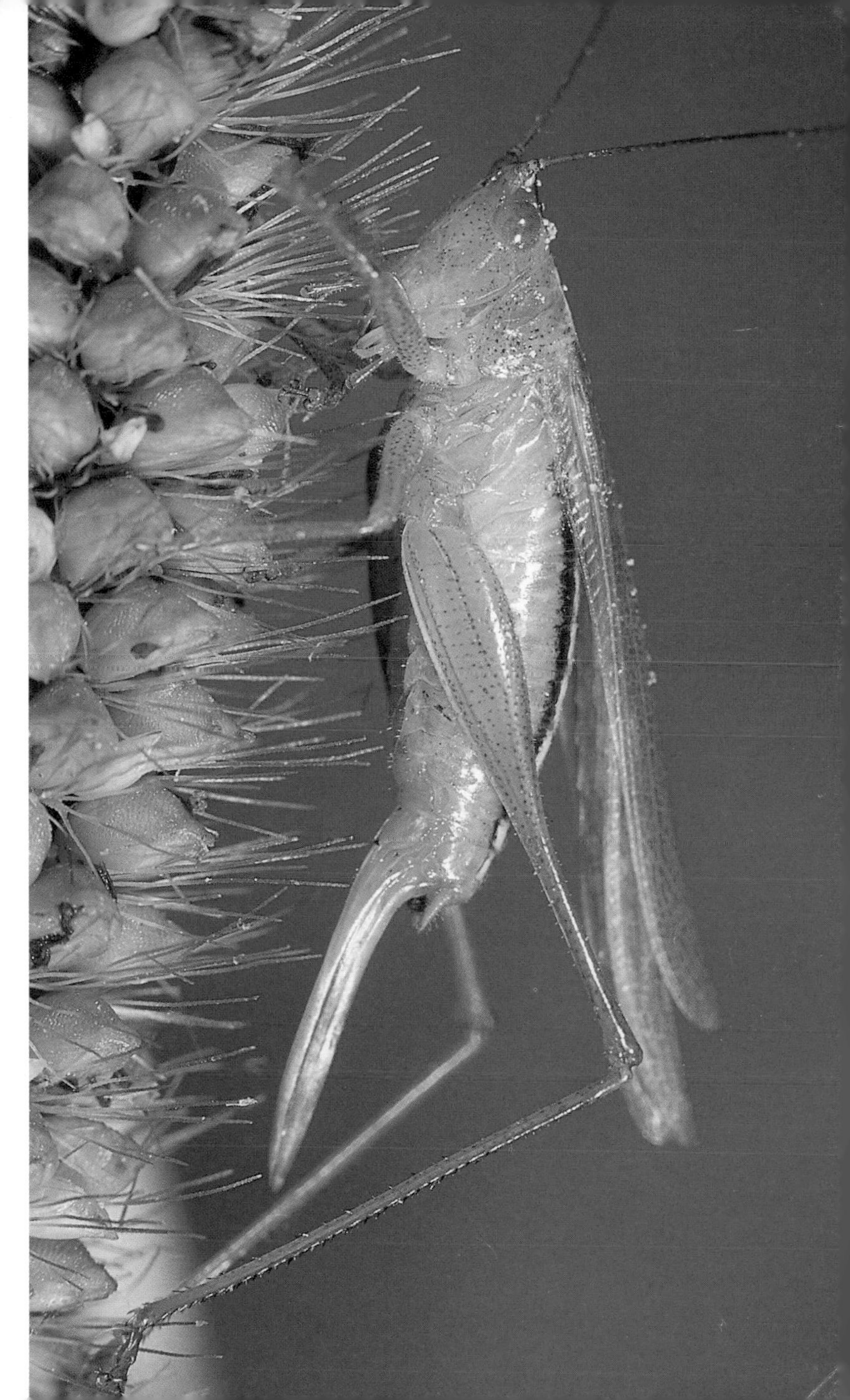

Above: **A slant-faced grasshopper**
Right: **A slender female meadow grasshopper (or slender meadow katydid)**

Grasshoppers, as you might guess from their name, live mainly in grasslands. Most are green or brown, to match the grass that surrounds them, but some have black, yellow, or red markings. Grasshoppers can be as much as 3 inches (7.5 cm) long, but most are smaller. In many species males are smaller than females.

"True" locusts belong to the family of short-horned grasshoppers.

Many common North American grasshoppers belong to a group called short-horned grasshoppers. They are active during the day and rest during the cool of the night. Locusts belong to this family. Long-horned grasshoppers—so called because of their long antennae—form a different group. This family includes the katydids, of which there are several types. Katydids, like many members of this family, are most active in the early evenings.

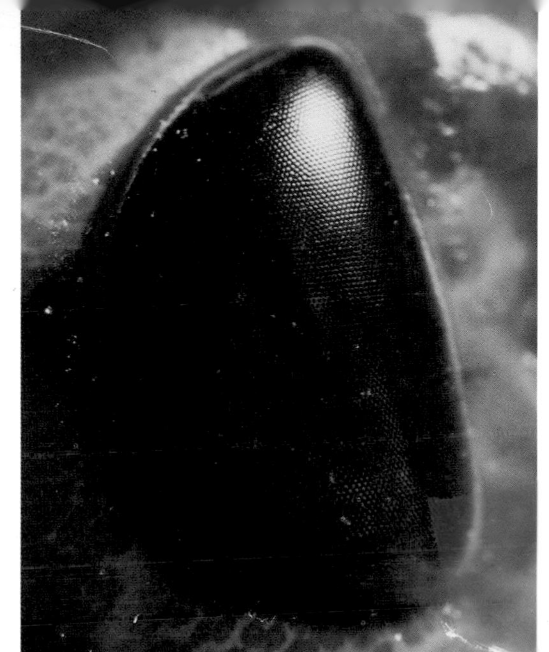

The large black compound eyes of this short-horned grasshopper are actually made up of thousands of tiny lenses.

A CLOSER LOOK

Like all insects, crickets and grasshoppers have six legs and three main body parts: head, thorax, and abdomen. Like all insects they also have a hard shell on the outside—an exoskeleton—instead of an internal skeleton.

Face to face, these insects look like something from a science-fiction film. A cricket or grasshopper has five eyes. Two large compound eyes, with many tiny lenses, detect objects and movement. Three other eyes—one between the compound eyes, and one on each side—respond to light. They are so small that they are hard to see.

Crickets and grasshoppers have five eyes, though only the big compound eyes can be easily seen.

Sensitive antennae help the insects find food, which they bite and chew with powerful jaws that move side to side. They taste their food with palpi, organs located alongside the mouth. Grasshoppers mostly eat plants—leaves, flower petals, seeds—and they are big eaters. In swarms, they can destroy crops and have even been known to chew up laundry hung outside to dry. A few types of grasshoppers eat other insects. Crickets also eat plants and just about anything else—even wool clothing, and, occasionally, other crickets.

Grasshoppers and crickets have powerful jaws that they use to chew leaves, flowers, and seeds.

"EARS" ON THE LEGS

The "ears" of crickets and many grasshoppers are located on the insects' legs or abdomen! These "ears" are two small membranes called tympanal organs. They are on the front legs of crickets and long-horned grasshoppers, and on the abdomen of short-horned grasshoppers.

The tympanal organ of a cricket or grasshopper is very similar to a human eardrum.

Breathing holes, or spiracles, are located on the thorax and abdomen. Crickets have two sensitive spikes at the end of the abdomen. These spikes may help the cricket find its way when it scuttles backwards. Female grasshoppers and crickets also have a stalk-like organ called an ovipositor at the end of the abdomen. They use this organ to lay eggs.

Females lay their eggs by using stalk-like organs called ovipositors.
Inset top: **Most spiracles, or breathing holes, are located on a grasshopper's or cricket's abdomen and thorax.**
Inset bottom: **A female plants her ovipositor in sandy soil to deposit her eggs.**

HOPPERS AND SINGERS

Most adult grasshoppers and crickets have two pairs of wings attached at the thorax. Grasshoppers can fly, but they can also leap amazing distances—up to 20 times their body length—with their powerful hind legs. Crickets don't fly. They use their wings only for making music.

Crickets and grasshoppers "sing" by rubbing parts of their body together. This is called stridulation. Crickets and long-horned grasshoppers rub their wings together. They draw the rough edge of a front wing, called the file, across a special ridge, or scraper, on the other front wing. Some short-horned grasshoppers rub a front wing over a hind wing. Others rub a hind leg over a front wing.

The rough surfaces of hind legs or wings create "music" when they are rubbed together.

A short-horned grasshopper leaps effortlessly through the air.

A CRICKET THERMOMETER

Like other insects, crickets and grasshoppers are more active in warm temperatures—and they chirp more when it's warm, too. You can even guess the temperature by listening to the chirps of crickets. If field crickets fall silent on a cool evening, the temperature has probably dropped below 55 degrees F (12.8 degrees C).

If you hear snowy tree crickets, you can get a more precise temperature reading. These crickets have a distinctive call—a rapid treet-treet-treet, varying loud and soft. The warmer the air, the faster they chirp. To find out the temperature on the Fahrenheit scale, count the number of chirps you hear in 15 seconds, then add 40.

Short-horned grasshoppers are heard mainly during the day. Although they often sing at dusk and on cloudy days, crickets and katydids mostly sing at night. Each species has a distinctive call. The true katydid is named for its song, which sounds like *katy-did, katy-didn't*—repeated as many as 60 times a minute.

With very few exceptions, only males "sing." They call to let others of their kind know where they are and to warn other males to stay out of their territory. Male crickets are especially aggressive—they will fight rivals who don't take their chirped warning. The main reason that crickets and grasshoppers sing is to attract mates. That is why you are most likely to hear these insects chirping away during their mating season, late in the summer and early in the fall.

FROM EGG TO ADULT

About a week after mating, the female short-horned grasshopper digs down with her ovipositor an inch (2.5 cm) or more in the ground, and lays from 20 to 120 eggs. She surrounds them with foam that hardens into a protective egg pod. Field crickets also lay their eggs in the ground, usually one at a time rather than in pods. Mole crickets lay eggs in chambers as much as 8 inches (20 cm) underground. Snowy tree crickets lay eggs inside plant stems. Katydids and many other long-horned grasshoppers also lay eggs in plant tissues.

***Top:* A slender meadow katydid lays her eggs in a plant stem.**
***Bottom:* A two-striped short-horned grasshopper lays her eggs in soil.**

Some grasshoppers surround their eggs with a protective foam that hardens into an egg case.

A green-striped grasshopper nymph

The discarded skin of a nymph that has molted

Crickets and grasshoppers generally don't live through winter in cold climates, but some eggs laid in the fall survive. Tiny new insects, or nymphs, hatch out in the spring. They look like little adults without wings. They can't fly, but they can jump.

The nymphs feed on plants, eating up to twice their body weight each day. They grow fast. But the outer skeleton cannot grow, so the nymph sheds its skin, or molts, as often as once a week. The nymph breaks through the skin with a hard area called the ampulla, at the back of its head. Then it just walks out of the old skin. A new skin has already formed, and it quickly dries into a hard shell.

After four to six molts, most crickets and grasshoppers are adult, with fully developed wings. The wings are soft, at first, but they dry hard within an hour or two of the insect's final molt. Grasshoppers are then ready to fly, and both crickets and grasshoppers can begin to sing.

Crickets and grasshoppers have many natural enemies. Spiders, mantises, birds, lizards, and snakes are among the animals that dine on them. The insects' main defense from predators is to hop, scuttle, or fly away. Some grasshoppers spit a dark liquid when they're alarmed. This startles predators and gives the grasshopper a chance to escape. But even if they escape predators, adult crickets and grasshoppers generally live only a few months—just long enough to mate and keep their reproduction cycle of life going.

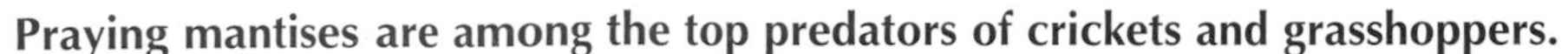

Praying mantises are among the top predators of crickets and grasshoppers.

A yellow and black garden spider wraps up its grasshopper prey.

A MORMON MIRACLE

When Mormons first settled in Utah, they had barely enough food to get through the first winter. Their survival depended on their first harvest. And just when it seemed that the harvest would be good, swarms of grasshoppers descended on the crops. The settlers tried to beat the insects off the crops, to drown them, to burn them—all without success. Then, to the settlers' amazement, flocks of sea gulls began to fly in from the west. The gulls swooped down and devoured the grasshoppers, saving enough of the harvest to let the settlement survive. The Mormons saw the arrival of the gulls as a miracle. Today a memorial to the gulls stands in Salt Lake City, and the sea gull is the Utah state bird.

GRASSHOPPERS, CRICKETS, AND PEOPLE

Perhaps it's just as well that crickets and grasshoppers have so many natural enemies. In large numbers, these insects can cause big problems for humans.

Almost since people began to plant and harvest crops, they have been at war with grasshoppers. Swarms of grasshoppers, especially migrating locusts, can strip fields bare. And these insects can form huge bands, as flying adults or as swarms of nymphs marching and hopping across the ground. In the United States in the 1800s, farmers reported a band of migrating nymphs 23 miles (37 km) wide and 70 miles (113 km) long! Pesticides have controlled grasshoppers since then. But pesticides can harm the environment, and some of the most harmful ones, such as DDT, have been banned. Thus swarms of grasshoppers are still a problem for farmers in many parts of the world.

Above: **House crickets can be noisy, but according to folklore, they are also good luck.**
Right: **In large numbers, grasshoppers that eat crops pose a serious problem to farmers.**

Crickets can also be pests, especially in large numbers. Even one loose cricket can be annoying in a house, chirping through the night. But crickets are also helpful; they eat decaying plant material and, at times, other insects. In Asia crickets are sometimes kept as pets and placed in beautiful little cages. Varieties with especially musical chirps are highly prized.

Crickets are also said to bring good luck. That idea is the basis of a famous story by the English writer Charles Dickens, *The Cricket on the Hearth*. In it, a house cricket sings a "fireside song of comfort" to herald good times for a family, and falls silent during hard times.

Keeping Crickets and Grasshoppers

You are more likely to hear crickets and grasshoppers than you are to see them. They blend in well with their surroundings, and they move quickly. That makes it difficult to observe these insects in the wild. But it is not so difficult to capture crickets and grasshoppers. Most kinds can be kept for a short time, if you provide the food and living conditions they need.

In this section, you will find instructions for collecting crickets and grasshoppers in the wild and for caring for them at home. Crickets can be bought at bait shops and at most pet stores, which keep a supply of these insects as food for lizards and similar pets. These insects are also available through the mail-order sources listed on page 46.

After you have finished watching your insects, release the ones that you collected in the wild. Bring them back to the place where you found them, or to a similar spot. Do not release insects that you buy, however. If they are not naturally found in your area, they may not survive—or they may survive to become pests.

A field cricket

COLLECTION

Late summer and early fall are the best times to collect crickets and grasshoppers. The types you find will depend on where you live. Check in your local library for a field guide to insects, which will help you identify and learn more about them. Crickets and short-horned grasshoppers do best in captivity. Some katydids have special food needs, so it is best not to collect them.

Look for crickets at dusk, on cloudy days, or whenever you hear them singing. Follow the sound of their calls to find them. They will fall silent when they hear you nearby, but if you wait quietly they will chirp again. Field crickets are the easiest type to locate and capture. Look for them on the ground. They often hide among dead leaves and under logs and rocks. You may even find them in your house or garage.

Short-horned grasshoppers like this one do best in captivity.

If you're quick, you can catch crickets with your hands, or by popping a can or other container over them. Cup your hands so that you don't harm the insects, and put them in a small jar or a coffee can, with holes punched in the top to let in air. If it will be a while before you can set up a more permanent home for them, put in a few leaves or some grass for food. Keep the container out of direct sunlight.

Look for grasshoppers on warm, sunny days. You may see them hopping or flying over meadows, but they are hard to spot—they move very fast, and they are masters of camouflage. They are much harder than crickets to catch with your hands. And some types may spit a dark liquid or even bite if you do get your hands on them.

If alarmed, some grasshoppers will spit a black liquid at a potential enemy.

What You Need:

* A mop or broom handle
* Duct tape or black electrical tape
* A piece of closely woven fabric or muslind
* A wire clothes hanger
* Needle and thread

One easy way to catch crickets or grasshoppers is to sweep a net through tall grass in an area where you have seen or heard these insects. You can buy a collecting net from sources such as those listed on page 46, or you can make one.

What to Do:

1. Bend the clothes hanger into a round loop. Straighten the hanger's hook. (Ask an adult to help if the hanger is hard to bend.)
2. If you're using closely woven fabric or muslin, fold the material in half. Sew two of the three open edges closed, to form a deep pouch.
3. Put your pouch through the wire loop. At the open end, fold the fabric back over the wire, and sew it in place.
4. Tape the net to the handle.

CARE

If you are going to keep your insects longer than a day or so, give them a special home. You can use a small aquarium or a large jar. Make sure that the container has a secure top—you won't want crickets or grasshoppers to get loose in your house. Use a lid with holes punched in it for ventilation, or cover the container with mesh, secured by a rubber band.

A cricket home

Sections of egg cartons provide good resting places for crickets.

Place some loose, dry sand in the bottom of the container. If you are keeping crickets, give them some places to hide. Sections of egg carton work well. Keep the container out of direct sunlight.

Make sure the insects have fresh food and water every day. Grasshoppers and field crickets will eat grasses, green leaves, and cut-up pieces of fruit. Lettuce leaves provide moisture as well as food, and wheat bran gives the insects some roughage. Crickets may also eat small pieces of breakfast cereal, meat, or dog biscuits. They seem to need the protein contained in these foods, and if they have an adequate supply, they may be less likely to fight with each other.

Remove old food and add fresh items daily, and remove insect waste as it builds up. Provide water by misting the inside of the container with a spray bottle. Or put in cotton balls or pieces of clean sponge that have been soaked in water. You can also put water in a small bottle top and place it in your insect home. Do not use a deeper water dish, though, because your insects might drown in it.

3

Investigating Crickets and Grasshoppers

You can learn a great deal about crickets and grasshoppers just by watching them up close in the home you have provided for them. In this section, you will find some activities that will help you learn even more. Many of these activities can be done with grasshoppers or crickets. Crickets make better subjects because they are easier to handle—especially when the insects must be moved from one container to another. Since they can't fly and don't leap as far as grasshoppers, crickets are less likely to escape. And activities with crickets can be done at anytime because they are available year-round at pet and bait shops and through mail-order sources.

Have fun with these activities. Remember to always handle the insects gently, and to return those you collected in the wild to the place where you found them.

DO CRICKETS LIKE TO LIVE TOGETHER?

Crickets often chirp in groups, calling back and forth to each other. Do the crickets actually like to live together, or do they prefer solitary living quarters? Make a prediction based on what you have read about these insects. Then find out if you are right by doing this activity.

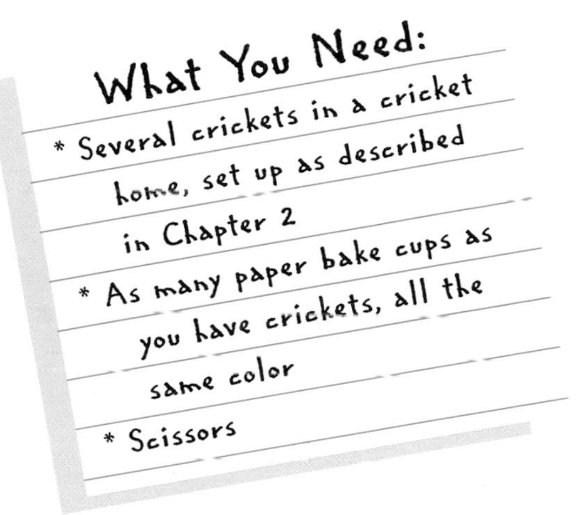

What to Do:

1. Cut a small "doorway" from the rim of each paper bake cup.
2. Place the cups upside down in the cricket home. Crickets will tend to go inside the cups to hide.

3. Check the cups from time to time, lifting each one to see how many crickets are inside.

Results: Each time you check the cups, write down the number of crickets you find under each one. Repeat the activity, moving the cups to different locations in the cricket home.

Conclusion: Did your crickets crowd in together, or did they prefer "single rooms"? What do your results suggest about the way crickets live in the wild?

WHAT NATURAL SURROUNDINGS DO CRICKETS PREFER?

Crickets are easy-going insects that make their homes in many places. Field crickets are especially likely to turn up anywhere. Given a choice, what surroundings do you think your insects would prefer? Make a guess, and then do this activity.

What You Need:

* A small aquarium or similar flat-boltomed container, to use as a temporary cricket home
* Mesh and a rubber band, to cover the container
* A variety of natural materials—dry sand, damp sand, loose soil, grass clippings, leaves
* Several crickets

What to Do:

1. Place your natural materials in the container, putting each in a different spot. Don't put food or water in the container.
2. Put crickets in the container, and cover it securely. Watch for a while to see where the crickets go.

3. Return the crickets to their regular home after an hour or so. They will need to have some food and water.

Results: The crickets may explore their temporary home a bit, but then they will probably settle in one area. Note which materials they prefer. Repeat the experiment using other materials, such as gravel and pieces of bark.

Conclusion: What do your results tell you about the places where you would be likely to find crickets in the wild?

DO CRICKETS PREFER CERTAIN COLORS?

Crickets and grasshoppers are hard to see in the wild because they blend in so well with their surroundings. Do they seek out certain colors? Decide what you think, and then do this experiment to see if you are right.

What to Do:

1. Put the blocks in the cricket home.
2. Crickets like to climb, so they will probably climb on and around the blocks. Check every 15 minutes or so to see where the crickets are.

3. Change the experiment—move the blocks, or put in different colors.

Results: Each time you check the container, write down how many crickets are on or touching each block.

Conclusion: Which blocks did the crickets prefer? You can try this activity using blocks colored black, white, and shades of gray, to see if they prefer darker or lighter shades.

WHAT MAKES CRICKETS SING?

At times, crickets chirp loudly—and frequently, they chirp hardly at all. What factors do you think inspire these little music makers? Answer based on what you know about crickets, and then do this activity. It works best if you have lots of crickets—males, females, immature, and adult. But you do not need all these to do a limited version.

What You Need:

* A number of crickets
* Small containers, such as baby food jars, all the same size
* Mesh and rubber band to cover the jars
* Sand, cricket food, and moistened cotton balls

What to Do:

1. Set up each small jar as a cricket home. Put some slightly moistened sand on the bottom, and add food and a moistened cotton ball for water. Each cricket home should be identical.

2. Put crickets in the homes, alone and in various pairings. Here are some possibilities:
   Jar 1—Immature female (this cricket has a long "tail"—her ovipositor—but no wings)
   Jar 2—Immature male (no ovipositor, no wings)
   Jar 3—Adult female (ovipositor, wings)
   Jar 4—Adult male (wings, but no ovipositor)
   Jar 5—Two adult males
   Jar 6—One adult male and one adult female
   You can vary this, and make other pairings, depending on the crickets you have.
3. Place the containers in a warm place, out of direct sunlight. Space them out, so you'll be able to tell which crickets chirp. But be sure all the jars receive the same amount of light.

Immature male cricket (small wing pods)

Mature male cricket (full-size wings)

Immature female cricket (small wing pods, ovipositor)

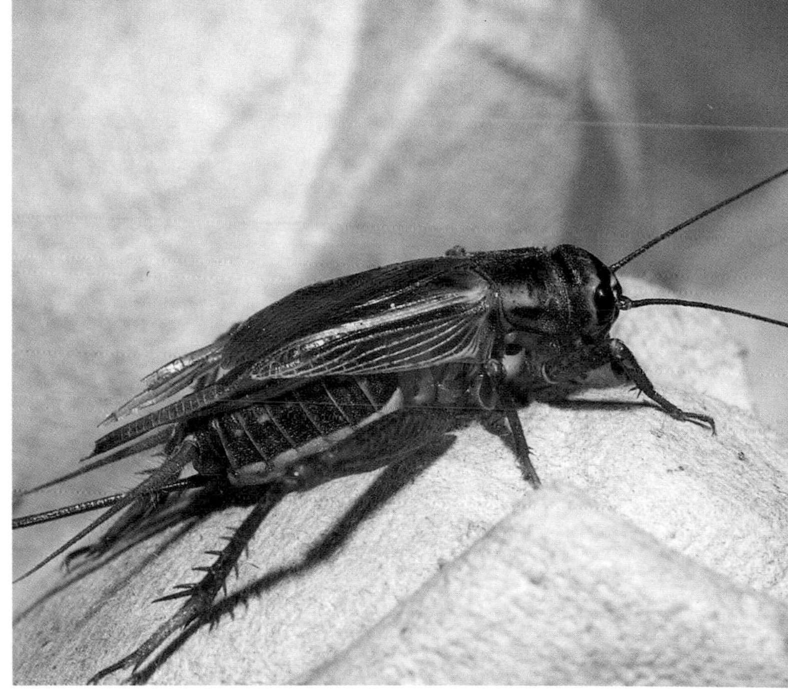

Mature female cricket (full-size wings and ovipositor)

4. Each time you hear chirping, sneak up quietly to find out which cricket is making noise. (The cricket will stop chirping if it hears or sees you.) Write down your observation.

5. Repeat the experiment, moving all the jars to a different location. Try a room with bright light, a dark room, or a cool place, such as a basement.

Results: Which crickets chirped most?
Conclusion: What do your results tell you about the factors that affect chirping?

MORE ACTIVITIES WITH CRICKETS AND GRASSHOPPERS

1. Are your crickets lefties or righties? Adult male crickets sing by rubbing one wing over the other wing. Some rub the left wing over the right; others, the right wing over the left—but a cricket always rubs with the same wing. Watch your crickets to see which they use.
2. If you have immature crickets or grasshoppers (without wings), you may have a chance to see how they molt. Insects usually molt in late morning, when the warm sun will help dry and harden their new skin. Watch the molting and write down your observations. What time did molting begin? How long did the insect take to shed the old skin? How long did it stay still afterward? What color was the new skin? Did the color change as it dried?
3. If you have both male and female adults, they may mate, and the females may lay eggs. Because captive crickets and grasshoppers often eat eggs in their container, you'll need to provide a special place for the eggs. Fill a small container with damp sand, which the females prefer for egg-laying. Put a piece of fine wire screening over the top, and place it in your insect home. Push the container down into the dry sand on the bottom, or provide a ramp for the insects to climb up. The females will be able to lay eggs through the screening, and the eggs will be safe beneath it.

Adult male crickets sing by rubbing one wing over the other.

RESULTS AND CONCLUSIONS

Here are some possible results and conclusions for the activities on pages 33 to 41. Many factors may affect the outcomes of these activities—the types of insects you use, the type of home you set up for them, and other conditions. If your results differ, try to think of reasons why. Repeat the activity with different conditions, and see if your results change.

Do Crickets Like to Live Together?

Your results may vary, but you'll probably find a cricket under each cup. Although they sing to each other, most crickets are solitary and prefer to live alone. However, a singing male may attract females to his cup.

What Natural Surroundings Do Crickets Prefer?

Your results will depend on the type of insects you have and on the materials you provide. Most crickets will seek out materials such as leaves, where they can hide.

Do Crickets Prefer Certain Colors?

Your results will depend on the variety of the colors you provide. Our crickets preferred a natural-colored block. It was closest to their own color and to the color of dried leaves and grasses, which they might find in nature.

What Makes Crickets Sing?

Only adult males sing, so you won't hear chirping from jars that do not have an adult male. Males usually chirp more in the presence of a female. They chirp aggressively when another male is present, and the two males may fight. (Two females may also fight.) Warmth and darkness tend to increase chirping.

SOME WORDS ABOUT CRICKETS AND GRASSHOPPERS

ampulla A hard region at the back of the head, which immature crickets and grasshoppers use to break out of their eggs and out of their skins when they molt.

camouflage Coloring that blends in with the surroundings.

exoskeleton The hard outer skin of an insect. It takes the place of an internal skeleton, protecting the insect and providing attachment points for muscles.

molt To shed an old exoskeleton that has become too small.

nymph An immature cricket or grasshopper.

ovipositor A tube, extending from the back of the abdomen, that female crickets and grasshoppers use to lay eggs.

predators Animals that catch and eat other animals.

solitary Living or being alone.

spiracles Holes through which crickets and grasshoppers breathe.

stridulation Making noise by rubbing body parts together.

SOURCES FOR CRICKET AND GRASSHOPPER SUPPLIES

You can buy crickets at many pet stores and bait shops. The companies below sell crickets, collecting nets, and other supplies through the mail. If you obtain insects through male order sources such as these, do not release them into the wild.

Carolina Biological Supply
2700 York Road
Burlington, NC 27215
1-800-334-5551

Connecticut Valley Biological
82 Valley Rad, P.O. Box 326
Southampton, MA 01073
1-800-628-7748

FURTHER READING

Dallinger, Jane. *Grasshoppers.* Minneapolis, MN: Lerner, 1990.
Johnson, Sylvia. *Chirping Insects.* Minneapolis, MN: Lerner, 1990.
Kubo, Hidekazu. *The Grasshopper.* Raintree, Steck Vaughn, 1990.
Lampson, Christopher. *Insect Attack.* Brookfield, CT: Millbrook, 1992.
Perry, Phyllis, J. *Fiddlehoppers: Crickets, Katydids, and Locusts.* New York: Franklin Watts, 1995.

INDEX

Note: Page numbers in italics indicate pictures.

Photo Credits

Page 22: © Salt Lake Convention and Visitors Bureau